W9-DEH-304

Geometry at Every Turn

Rachel Chappell

Rourke
Publishing LLC
Vero Beach, Florida 32964

www.rourkepublishing.com

PHOTO CREDITS: © Kevin Russ: page 8; © Pierangelo Rendina: page 14; © Andresr: page 15; © Aaron Kohr: page 19

Editor: Robert Stengard-Olliges

Cover design by Nicola Stratford.

Library of Congress Cataloging-in-Publication Data

Chappell, Rachel M., 1978-
  Geometry at every turn / Rachel Chappell.
    p. cm. -- (My first math)
  ISBN 1-59515-974-6 (hardcover)
  ISBN 1-59515-944-4 (paperback)
 1. Geometry--Juvenile literature. 2. Shapes--Juvenile literature. I.
Title.
  QA445.5.C53 2007
  516--dc22

                    2006019781

Printed in the USA

CG/CG

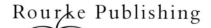

Rourke Publishing

www.rourkepublishing.com – sales@rourkepublishing.com
Post Office Box 3328, Vero Beach, FL 32964

Table of Contents

Geometry is Everywhere

Everywhere we look we can see lines and shapes. Lines form shapes. A STOP sign has a shape. A red light has a different shape. When you see lines and shapes you are observing the **geometry** in your world.

Lines and shapes, are in everything we see and touch. Toys, packages, playground equipment, buildings, clothes! Even food!

Lines

Some lines are curved. They form some of the shapes we see. They form circles and ovals. And don't forget rainbows!

Some lines are straight. They go from one place to another. The wires on a bridge stretch straight down from its top to its bottom.

Lines can be **parallel**. That means they are the same distance apart as they go from place to place. Two parallel straight lines look like railroad tracks. Parallel curved lines look like the strips of color in a rainbow.

The edges of a sidewalk run parallel.

Angles

Sometimes two lines cross each other. They **intersect**. When they do, they form an angle.

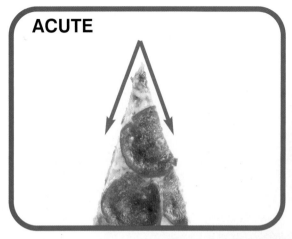

ACUTE

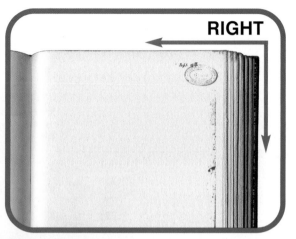

RIGHT

OBTUSE

Small angles like the ones at the point of a piece of pizza are called **acute angles**. Angles like the corner of the page of this book are called **right angles**. Angles that are bigger than that are called **obtuse angles**.

Can you find the acute, right, and obtuse angles on the playground?

When you look at a house you can see how the lines of the roof and sides form angles. If the roof is steep, the lines may form an acute angle. Can you find the right angles at the corners of the windows and doors?

Polygons

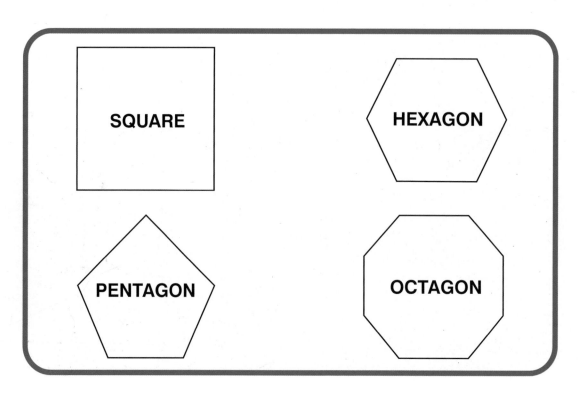

Straight lines can make shapes too. Straight lines can zigzag all over the place or form a closed shape. The closed shapes are called **polygons**. You can sort polygons into groups by how many sides they have.

Solid Figures

If you look in your kitchen, you will find boxes, cans, and food of different shapes.

Boxes and cans are solid geometric forms. They are not like shapes on a flat surface. They are **three dimensional**. You can fill them up with stuff.

Look around. See the many lines, angles, and shapes. Geometry is at every turn.

Glossary

acute angle (uh KYOOT ANG guhl) — an angle smaller than a right angle

geometry (jee OM uh tree) — the relationship of points, lines, angles, surfaces, and solids in math

intersect (in tur SEKT) — to cross each other

obtuse angle (uhb TUSE ANG guhl) — an angle larger than a right angle

parallel (PA ruh lel) — lines that are side by side and the same distance apart

polygon (POL ee gon) — a closed plane figure bounded by three or more straight lines

right angle (RITE ANG guhl) — an angle formed by the perpendicular inersection of two straight lines; an angle of 90°

three dimensional (THREE duh MEN shuhn el) — any of the animals that make mother's milk and grow hair

Index

Further Reading

Fitzgerald, Teresa. *Math Dictionary for Kids.* Prufrock Press, 2006.

Keoke, Emory Dean. *American Indian Contributions to the World.* Facts On File, 2005.

Neuschwander, Cindy. *Mummy Math: An Adventure in Geometry.* Henry Holt, 2005.

Websites To Visit

www.mathleague.com/help/geometry/geometry.html
www.math.okstate.edu/~rpsc/dict/Dictionary.html
www.coolmath4kids.com/

About The Author

Rachel M. Chappell graduated from the University of South Florida. She enjoys teaching children as well as their teachers. She lives in Sarasota, Florida and gets excited about reading and writing in her spare time. Her family includes her husband, one son and a dog named Sadie.

MY FIRST MATH

G9

Teaching Tool: Comprehension Questions
What Did You Learn?

1. Can you identify two objects that contain angles?
2. What is a polygon? How many polygons can you name?
3. How are polygons and solid figures different?

TITLES IN THIS SERIES:

ISBN 1-59515-944-4

90000

9 781595 159441

ROURKE CLASSROOM RESOURCES
The path to student success